EARTH SCIENCE—GEOLOGY Need to Know

SilverTip

Igneous Rocks

by Ruth Owen

Consultant: E. Calvin Alexander Jr., Professor Emeritus
Earth & Environmental Sciences
University of Minnesota, Minneapolis

BEARPORT PUBLISHING

Minneapolis, Minnesota

Credits

Cover and Title Page, © www.sandatlas.org/Shutterstock, © ZaZa Studio/Shutterstock, and © DanielFreyr/Shutterstock; 3, © www.sandatlas.org/Shutterstock; 5T, © Virrage Images/Shutterstock; 5M, © Kabardins photo/Shutterstock; 5B, © Dmitry Kalinovsky/Shutterstock; 7, © vvoe/Shutterstock; 8, © Mopic/Shutterstock; 9, © franco lucato/Shutterstock; 10, © Shutterstock; 11, © ChristineMinato/Shutterstock; 13T, © Designua/Shutterstock; 13B, © Designua/Shutterstock; 14, © Designua/Shutterstock; 15, © USGS/Alamy; 16, © vfpictures/Shutterstock; 17, © Willyam Bradberry/Shutterstock; 19, © Attila JANDI/Shutterstock; 21, © Attila JANDI/Shutterstock; 22, © Sean Pavone/Shutterstock; 23, © Ellen Bronstayn/Shutterstock; 25, © Patrick Poendl/Shutterstock; 27, © imageBROKER/Alamy; 28, © VectorMine/Shutterstock.

President: Jen Jenson
Director of Product Development: Spencer Brinker
Senior Editor: Allison Juda
Associate Editor: Charly Haley
Designer: Colin O'Dea

Library of Congress Cataloging-in-Publication Data

Names: Owen, Ruth, 1967- author.
Title: Igneous rocks / by Ruth Owen.
Description: Silvertip books. | Minneapolis, Minnesota : Bearport Publishing Company, 2022. | Series: Earth science–geology: need to know | Includes bibliographical references and index.
Identifiers: LCCN 2021026720 (print) | LCCN 2021026721 (ebook) | ISBN 9781636915777 (library binding) | ISBN 9781636915845 (paperback) | ISBN 9781636915913 (ebook)
Subjects: LCSH: Igneous rocks–Juvenile literature.
Classification: LCC QE461 .O935 2022 (print) | LCC QE461 (ebook) | DDC 552/.1–dc23
LC record available at https://lccn.loc.gov/2021026720
LC ebook record available at https://lccn.loc.gov/2021026721

For more information, write to Bearport Publishing, 5357 Penn Avenue South, Minneapolis, MN 55419.
Printed in the United States of America.

Contents

Earth's Rocks Rock

What do sidewalks, frying pans, and dentist tools have in common? They can all be made using igneous (IG-nee-uhs) rocks! These rocks have many uses. But they are all formed in the same ways.

There are three kinds of rocks on Earth. They are igneous, **sedimentary** (*sed*-uh-MEN-tur-ee), and **metamorphic** (*met*-uh-MOR-fik). Rocks are grouped based on how they form.

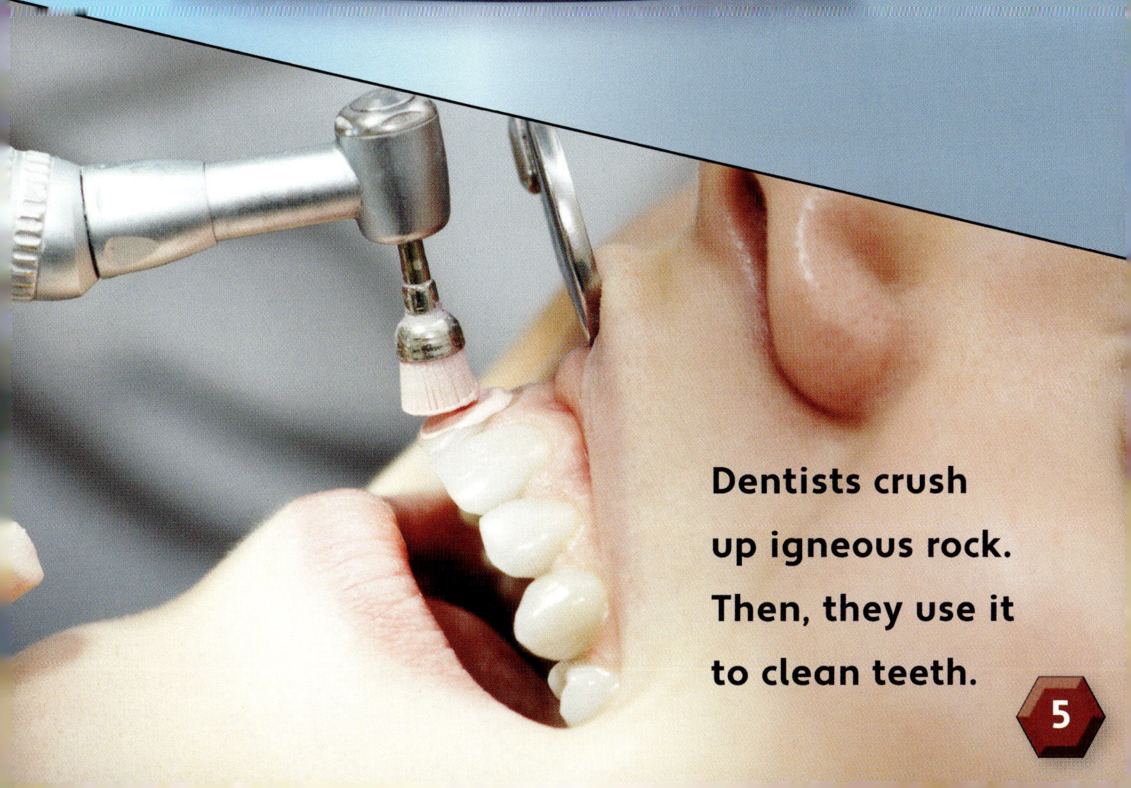

Dentists crush up igneous rock. Then, they use it to clean teeth.

Most rocks are made of **minerals**. The minerals in igneous rocks form as tiny **crystals**. Many of these small shapes lock together to make solid rock. Often, the crystals in igneous rock can only be seen with a **microscope**.

Some igneous rocks are made of just one mineral. Others are made from a mix of minerals.

Sometimes, you can see the differently colored minerals that come together in an igneous rock.

How Do Igneous Rocks Form?

The stuff that will become igneous rock starts inside our planet. Earth is covered with a crust made of rock. Just below that is the **mantle**. Here, it is so hot that rock becomes soft like caramel. The hot, flowing rock called **magma** can one day become igneous rocks.

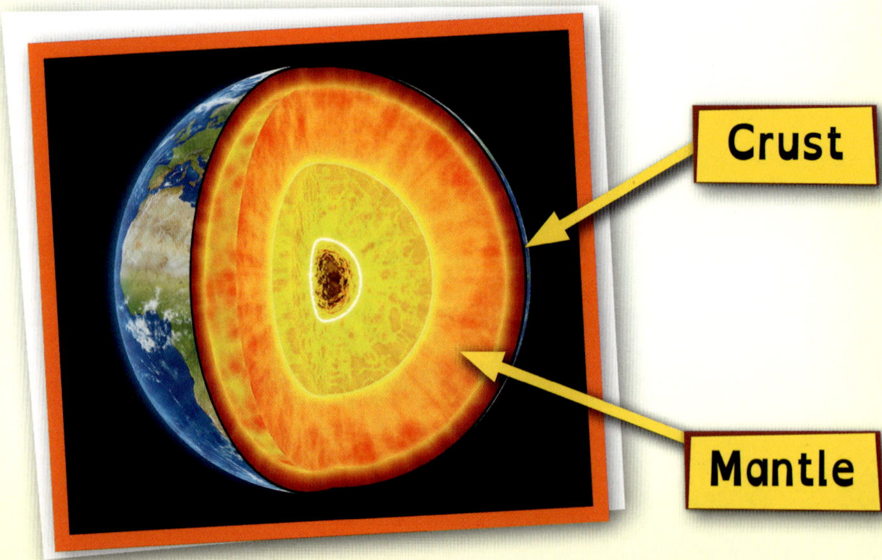

In many places, Earth's crust is covered with soil and plants. But sometimes we can see Earth's crust on the surface. Some mountains are tall points of Earth's crust.

Sometimes, Earth's crust cracks deep in the ground. Hot magma flows up into the crack. Once it is away from the heat of the mantle, the magma cools. Over thousands or millions of years, the magma gets hard. It becomes igneous rock.

Igneous rock that forms this way is often underground. People dig to find the rock. They use machines to cut it from the crust.

Igneous above Ground

Igneous rock can also form on Earth's surface. This happens when Earth's crust shifts.

Earth's crust is broken into giant pieces called **tectonic plates**. These plates are always moving. As they move, they can make cracks in the crust.

We don't feel Earth's tectonic plates moving because they are very slow. They move only about as fast as your fingernails grow.

Major Tectonic Plates

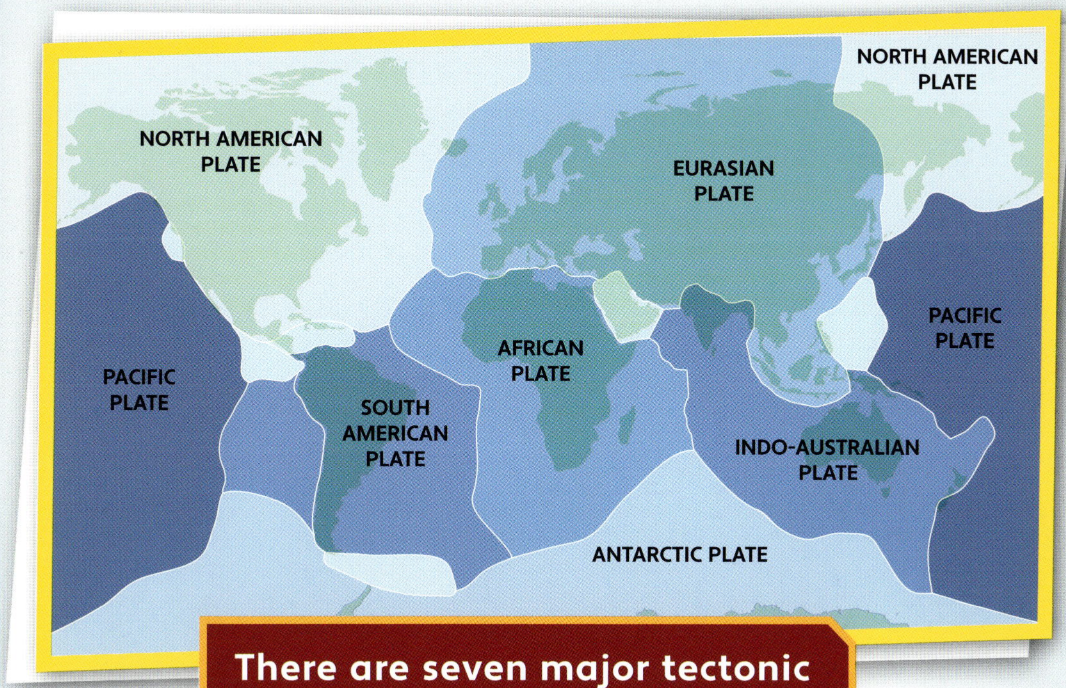

There are seven major tectonic plates and many smaller ones.

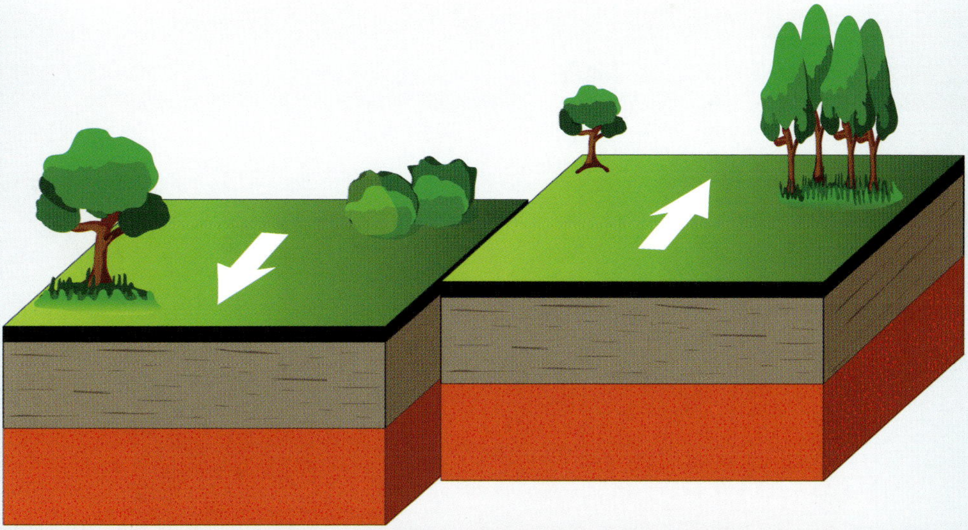

Tectonic plates can move in different ways.

Cracks in Earth's crust allows magma to flow out. Once magma makes its way onto Earth's surface, it's known as **lava**. The fiery lava spreads over land. As it cools, the lava gets thicker and slower. It starts to get harder.

A volcano is an opening in Earth's crust that lets out lava and steam.

Lava can be 2,000 degrees Fahrenheit (1,100 degrees Celsius). That's about 10 times hotter than boiling water!

Firming Up

Lava may be hard enough to walk on after just 15 minutes. But underneath the hard surface, the lava can stay hot and soft. It takes months or years to cool completely. But eventually it turns into solid igneous rock.

Hot gases can get trapped in lava. The gases form bubbles. When the bubbles pop, they leave behind tiny holes in the rock.

Magnificent Magma

Sometimes, lava cools into unusual shapes. Around 50 million years ago, hardening lava in Northern Ireland began to crack. The rock broke into tall columns with flat tops. The strange rocks came to be known as the Giant's Causeway.

The Giant's Causeway is made up of more than 40,000 columns. Each piece of igneous rock has straight sides.

The Giant's Causeway

Building Up and Washing Away

Some of the most well-known igneous rocks are **volcanic** mountains. Lava can flow from the same crack in Earth's crust again and again. The layers of cooled lava build up. Over time, a cone-shaped mountain is left behind.

It can take just a few years or many thousands of years for lava to build into a mountain. Mount St. Helens is a volcanic mountain in Washington. It took thousands of years to form.

Mount St. Helens blew off its pointed top during a huge explosion in 1980.

Sometimes, magma doesn't make it out of the volcano. Ship Rock in New Mexico formed inside a volcanic mountain. Millions of years ago, lava gathered in the mountain. It cooled and turned to igneous rock. Then, wind and rain slowly washed away the volcano. Now, all that's left is Ship Rock.

Ship Rock is 1,400 feet (430 m) tall. When the rock first formed, it was about 3,000 ft (914 m) under the ground.

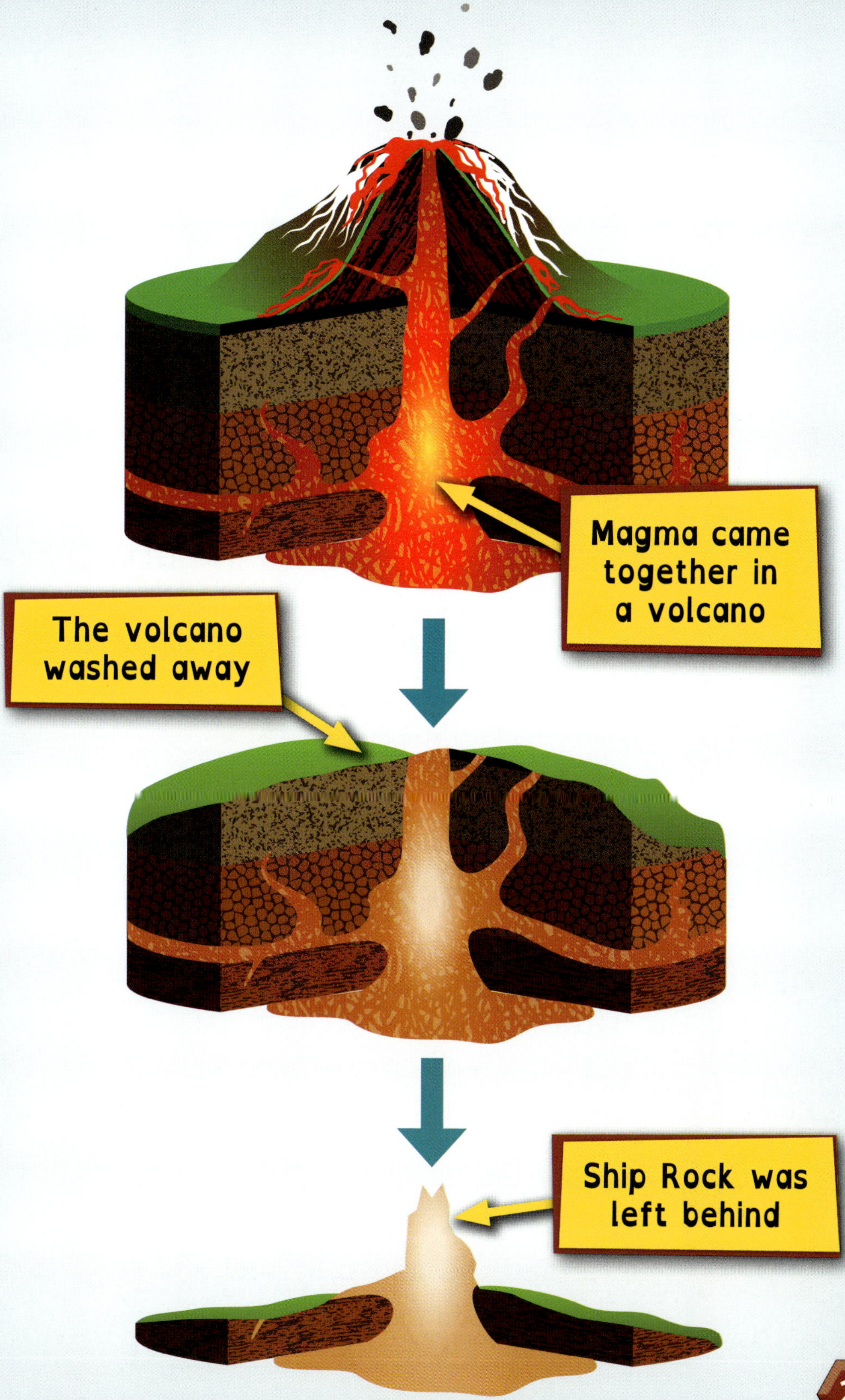
Magma came together in a volcano
The volcano washed away
Ship Rock was left behind

Uncovering Giants

California's Sierra Nevada Mountains were also once hidden by volcanoes. Millions of years ago, magma in Earth's crust hardened into igneous rock. Then, more magma flowed up to form volcanoes above the underground rock. Eventually, the volcanoes washed away. The Sierra Nevada Mountains were visible at last.

Once the igneous rock below the volcanoes was uncovered, weather shaped the rock. It formed the peaks of the Sierra Nevada Mountains.

The Sierra Nevada Mountains

A History of Earth

Igneous rocks shape our landscape. They are in our frying pans and front walks. But more than that, they tell a story. Thanks to igneous rocks, scientists know how Earth has changed. The rocks hold millions of years' worth of history. We can learn a lot when we look to rocks.

Scientists can date igneous rocks. Often, they look at how long things within the rock have been there. Knowing the age of igneous rocks can tell us what Earth looked like at different times.

Scientists who study rocks are called geologists.

A ROCKING WORLD:

Igneous Rocks

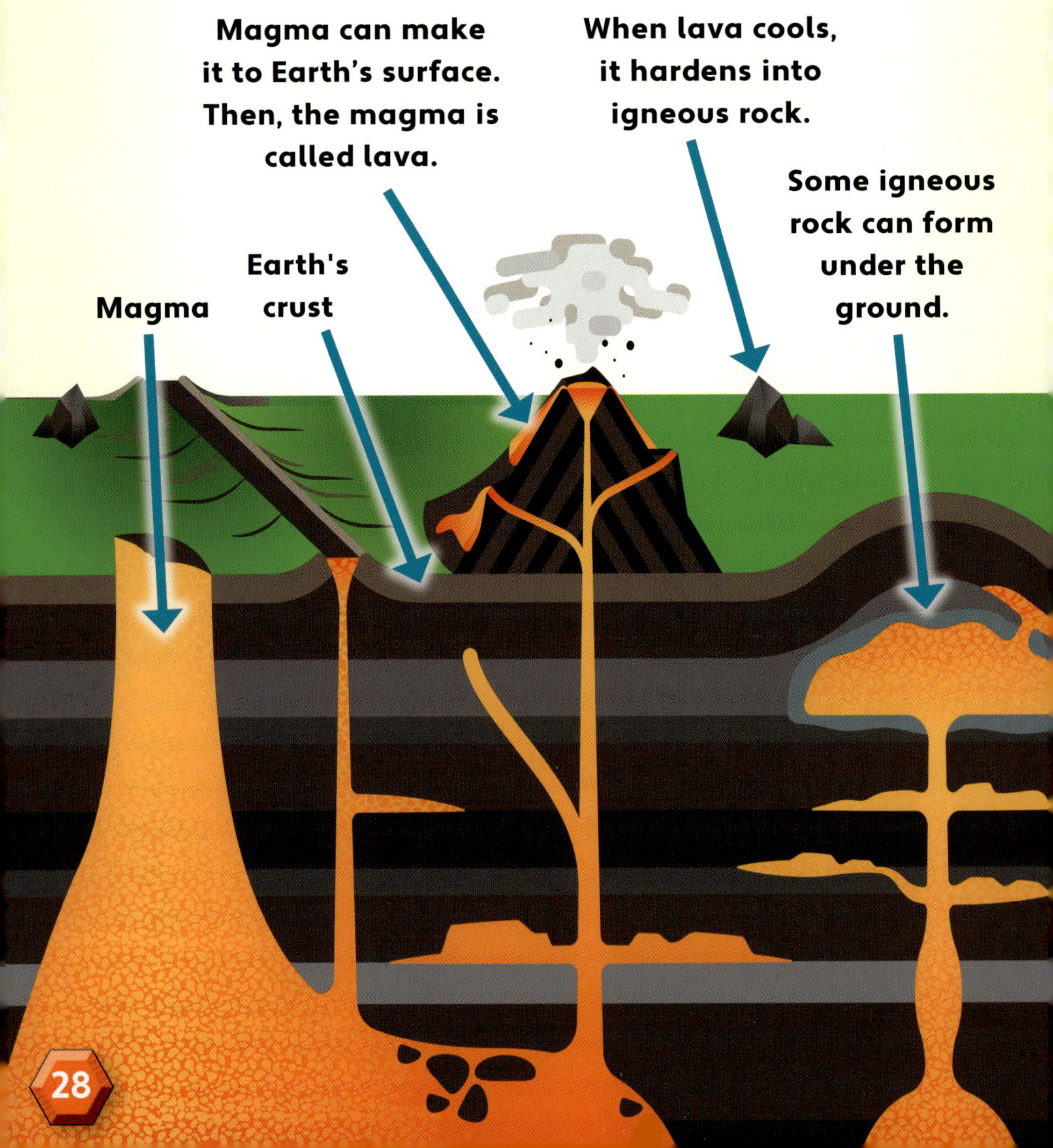

SilverTips for SUCCESS

★ SilverTips for REVIEW

Review what you've learned. Use the text to help you.

Define key terms

crust
magma
mantle
mineral
tectonic plates

Check for understanding

What are the three different kinds of rocks? How are they grouped?

Describe two ways igneous rocks are made. What is the same about these ways and what is different?

What role can volcanoes play in shaping igneous rocks?

Think deeper

What can scientists learn from studying igneous rocks? Why do these rocks matter?

★ SilverTips on TEST-TAKING

- **Make a study plan.** Ask your teacher what the test is going to cover. Then, set aside time to study a little bit every day.
- **Read all the questions carefully.** Be sure you know what is being asked.
- **Skip any questions** you don't know how to answer right away. Mark them and come back later if you have time.

Glossary

crystals the shapes in which minerals form

lava hot liquid rock that comes above ground

magma hot liquid rock flowing inside Earth

mantle the layer inside Earth just below the crust where magma flows

metamorphic rock that forms when rock changes from one type to another

microscope a tool that scientists use to see things that are too small to see with the eyes alone

minerals solid substances found in nature that make up rocks

sedimentary rock that forms from layers of tiny pieces of rock

tectonic plates sheets of rock that make up Earth's outer crust

volcanic of or made by a volcano

Read More

Fretland VanVoorst, Jenny. *Igneous Rocks (Rocks & Minerals).* Minneapolis: Bellwether Media, 2020.

Light, Charlie. *Volcanoes Reshape Earth! (Earth Under Construction).* New York: Gareth Stevens Publishing, 2021.

Pettiford, Rebecca. *Igneous Rocks (Geology Genius).* Minneapolis: Jump! 2019.

Learn More Online

1. Go to **www.factsurfer.com** or scan the QR code below.
2. Enter "**Igneous Rock**" into the search box.
3. Click on the cover of this book to see a list of websites.

Index

About the Author

Ruth Owen has written hundreds of nonfiction books. She lives on the rocky Cornish coast in England and has always been fascinated by rocks.